1/01

25.69

WIDE WORLD

PEOPLE *of the*
ISLANDS

Colm Regan

RAINTREE
STECK-VAUGHN
PUBLISHERS
A Steck-Vaughn Company

Austin, Texas

Island Life

Far out in the South Atlantic Ocean, midway between southern Africa and South America, lie the islands of Tristan da Cunha. Here, shipwrecked sailors once made their homes, building stone houses, gathering driftwood to make fires, and catching fish. From the shore, they could see only a vast and empty ocean. Descendants of those sailors still live on the islands today—the most isolated inhabited islands in the world.

The World's Largest Islands

Greenland	840,000 sq. mi. (2,175,600 sq. km)
New Guinea	317,000 sq. mi. (821,000 sq. km)
Borneo	287,400 sq. mi. (744,400 sq. km)
Madagascar	226,600 sq. mi. (587,000 sq. km)
Baffin Island	196,100 sq. mi. (508,000 sq. km)

The water that surrounds islands no longer cuts off island peoples in the way it once did. Ships and aircraft cross the oceans, carrying goods and people. Bridges and tunnels provide road and rail links. Telephones, television, and computers make communication quick and easy. Even the islanders of Tristan da Cunha have a radio station.

This boy lives on the ▶ island of Mindoro, in the Philippines.

4

In many ways, island life today is no different from life on the mainland. Yet some differences remain. Food and many other goods often have to be imported. Floods or volcanic eruptions can force people to abandon their island homes. When bad weather prevents an aircraft from landing with supplies, or an accident closes a vital bridge, islanders realize how much they depend on such links.

Isolation has some benefits too. On some islands, precious natural habitats remain almost untouched, and traditional customs and beliefs still survive. On islands where there are beautiful beaches and scenery, the tourist industry flourishes. It brings much-needed money to island peoples.

▲ Singapore is a densely populated island, crammed with high-rise buildings.

Island Animals
Some species of animals are unique to islands. There are no monkeys on the island of Madagascar, just off the African coast. Instead, monkeylike creatures called lemurs have developed. A great ape called the orangutan lives only on the islands of Borneo and Sumatra. Protecting the habitats of rare animals like these is an important task for island peoples.

▼ Many islanders earn their living by farming. These villagers live on a Fijian island in the Pacific Ocean.

Lands in the Ocean

The way in which islands formed has some impact on the people who live on them. Oceanic islands are formed by volcanoes under the sea. These pour out lava and ash, which gradually build up until they break through the surface of the water. The Hawaiian islands in the Pacific Ocean were formed in this way. The soil on volcanic islands is very rich, which has made them places where people want to settle and grow crops.

Sometimes, coral grows around a volcano, forming a reef. If the sea level drops, the reef stands up above the water, making a coral island. The soil of coral islands is usually thin and poor. Islanders have to rely on fishing as the main way of making a living.

▼ This map shows some of the world's major islands and groups of islands.

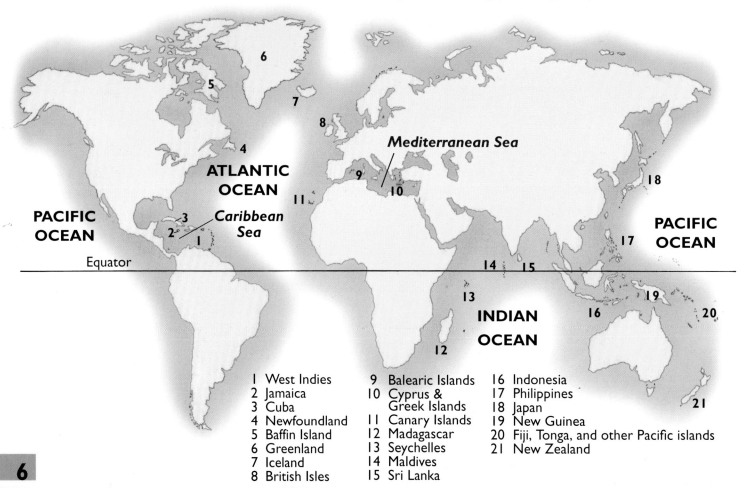

1	West Indies	9	Balearic Islands	16	Indonesia
2	Jamaica	10	Cyprus & Greek Islands	17	Philippines
3	Cuba			18	Japan
4	Newfoundland	11	Canary Islands	19	New Guinea
5	Baffin Island	12	Madagascar	20	Fiji, Tonga, and other Pacific islands
6	Greenland	13	Seychelles	21	New Zealand
7	Iceland	14	Maldives		
8	British Isles	15	Sri Lanka		

Hot Springs

More than 200 volcanoes have erupted in Iceland, and many of them are still active. Iceland also has more hot springs than any other country. That is because a fault in the earth's crust runs across the island. The islanders make the most of this natural hot-water supply by using it to heat their homes.

▲ Tourists watch as a hot spring gushes into the air at Strokkur, in Iceland.

Continental islands

Continental islands were once part of the mainland. Some, such as the island of Madagascar, which lies off the coast of Africa, became separated because of movements in the earth's crust. Others, such as the British Isles, were formed by changes in the sea level. Thousands of years ago, ice covered vast areas of land and sea. When the climate became warmer, the ice melted and the sea level rose. The low-lying areas of land that joined the British Isles to the continent of Europe were flooded, and the islands became separated.

Island climates

Island climates range from the ice and snow of Greenland, inside the Arctic Circle, to the tropical warmth of Borneo, which lies on the equator. Some islands or island groups are big enough to have a range of climates. The southern Japanese island of Kyushu has hot summers and mild winters, whereas the northern island of Hokkaido is much cooler, with snowy winters.

One feature that all islands have in common is that they are surrounded by water. Seas warm up and cool down more slowly than land. In the summer, this helps to keep island temperatures cooler than they are on nearby mainland. In the winter, island temperatures are warmer.

Tropical Storms

In the tropical regions of the world, ferocious storms develop far out at sea. In the Caribbean, these are called hurricanes, whereas in the Pacific they are known as typhoons. When they reach land, the heavy rain and strong winds they bring cause terrible damage, destroying buildings, flattening crops, and killing or injuring people. Islands can suffer particularly because they are often the first areas of land that the storms reach.

▼ Hurricane Hugo, in 1989, was one of the most serious storms ever to hit the Caribbean islands.

Island resources

Islands provide many different natural resources that people can use. These include the timber of tropical rain forests, plentiful supplies of fish, and some of the world's biggest mineral reserves. However, not all islands are equally rich in resources. In the Caribbean, for example, Trinidad and Jamaica have important mineral reserves but the islands of Antigua and Guadeloupe have almost none. This has affected the ways in which settlements and work on the islands have developed.

▼ Fish are an important natural resource. These islanders are preparing for a day's fishing off the coast of Sri Lanka.

History of Island Peoples

In the Philippines today, there are people who are descended from Indonesians, Malays, Chinese, Europeans, Americans, Indians, and Japanese. Many island populations are like this, made up of people from many different nations and races. In the past, islands were natural meeting places for travelers making long ocean journeys. They were also places that attracted settlers, who were eager to take over new land or to have a share in an island's rich resources.

Early settlers

The earliest islanders were people who crossed to continental islands when thin strips of land still joined them to the mainland. Many islands in the Pacific, for example, were linked to one another, or to Asia or Australia, 30,000 to 40,000 years ago. The Australian Aborigines and the Papuan peoples of New Guinea are among those who are descended from the first settlers.

▼ The traditions of some peoples in Papua New Guinea have lasted for centuries. These children are wearing traditional clothes and face decorations.

◀ These islanders from Tonga are paddling a traditional canoe, which has an outrigger to keep it stable. Similar boats were used by the earliest Pacific explorers.

Crossing the water

Later, travelers made short trips from the mainland to offshore islands in simple boats or on rafts. As their skills in boat building and navigation improved, they were able to make longer voyages. Early explorers, such as the Polynesians, used the chains of small volcanic islands in the Pacific as "stepping-stones" to travel farther and farther across the ocean. They were probably looking for more supplies of fish in the coral reefs around the islands.

A Home Away From Home

Iceland is an isolated island, with little fertile land, in the North Atlantic. It was settled just over 1,000 years ago by Vikings from Scandinavia. These people were already skilled at sailing the northern seas. They knew how to build houses to withstand the cold winters and how to find and grow food in difficult conditions. But very few other people have been tempted to settle in Iceland since those first travelers arrived.

▲ This seventeenth-century map was used by Dutch sailors. It shows islands and mainland that are now part of Malaysia, Indonesia, and the Philippines, in the Pacific Ocean.

Traveling to trade

As ancient peoples began to trade with one another, the movement of people from island to island increased. Traders were attracted to some islands because they were rich in natural resources. For example, 3,500 years ago, copper was mined on the island of Cyprus in the Mediterranean and sold to the island states of ancient Greece. Islands were also useful as stopping-off points on longer voyages, where traders could stock up on food and water or repair their boats.

As sailors met and traded with island peoples, they passed on their languages, customs, and religious beliefs. In the Indian Ocean, there are many clues to indicate which islands were visited by Arab traders. Some of the words in the Malagasy language of Madagascar are Arab words. The people of the Maldives became Muslims in the twelfth century, as a result of contact with Arab traders.

The search for the Spice Islands

In Europe, people had long relied upon Arab traders to bring them goods such as spices from the Far East. Spices were highly valued because they could be used to flavor food. As soon as Europeans were able to build ships that were strong enough to make long voyages, they set out to find routes to the Far East for themselves.

The Portuguese were the first Europeans to reach the famous Spice Islands—islands such as Java, Sumatra, and Sri Lanka. There, they were able to buy precious spices, such as cinnamon, cloves, and nutmeg. But the profitable trade in spices did not bring wealth to the islanders. Europeans fought local people and one another for control of the trade. In 1625, for example, the Dutch destroyed all the clove trees in the Moluccas except those on the island they controlled. This meant that all traders had to buy cloves from the Dutch.

▼ A sixteenth-century illustration of a Portuguese sailing ship, surrounded by flying fish. On ships like these, Portuguese explorers were able to reach the Spice Islands of the Far East.

The Maori

New Zealand was first settled by the Maori, who sailed there from Polynesia. In the 1800s, European settlers arrived, and the Maori lost their lands. Many were killed, and many more died from diseases brought to the country by the new settlers. After a century of conflict, attitudes have changed. Today, the Maori play an important part in the daily life and government of New Zealand.

◀ This woman, from Dominica, is one of the few descendants of the Carib people, who once lived in the West Indies. The Caribs were almost wiped out by Spanish settlers in the 1500s.

Island colonies

From the sixteenth century onward, many islands in the Caribbean and Pacific were taken over by Europeans. Most of these colonists, as they are called, came from Spain, Portugal, or Great Britain: for example, Cuba and the Philippines became colonies of Spain. For the colonists, the islands offered cheap or free land that was rich in natural resources. They began to grow crops such as sugarcane, coffee, and tobacco, which could be shipped back to their home countries and sold. Many colonists became wealthy. But the people who had been living on the islands before the colonists arrived lost their lands and many of them were killed.

Slavery

During the 1500s, the demand for sugar and tobacco grew in Europe. The colonists on the Caribbean islands needed more and more workers to grow the crops on their plantations. Over the next 300 years, more than 10 million people from Africa were forced to become slaves and taken to the Caribbean to work. When slavery was finally abolished, in the late nineteenth century, many slaves settled on the islands where they had worked. Their descendants still live there today. In Jamaica, for example, 75 percent of the population is of African origin.

Independence

During the twentieth century, many colonies have become independent. Islands in which local people govern themselves include Sri Lanka, Cuba, Jamaica, and the Philippines. Some islands, such as Montserrat in the Caribbean and St. Helena in the South Atlantic, are semi-independent. They have their own elected assemblies, but certain aspects of government, such as defense and foreign policy, are controlled by the nations that once ran the islands as colonies.

▲ This engraving, from 1833, shows slaves at work on Antigua in the West Indies. They are preparing the land so that sugarcane can be planted.

Work on Islands

Many islanders spend much of their working time producing food for their families. Subsistence farming, as this is called, usually takes place on islands that are too small, or cold, or where the soil is not rich enough for people to grow enough crops to sell for profit. In Haiti, a country on the Caribbean island of Hispaniola, almost two thirds of all workers farm this way.

Growing cash crops

On islands in the tropics, where the climate is warm and wet and the soil is rich, farming is big business. There, many islanders work on huge plantations to produce cash crops—crops that can be sold, usually abroad. Cash crops include coffee, tea, cocoa, tobacco, and bananas and other fruits.

▼ A woman picks tea leaves on a plantation in Sri Lanka.

Sugarcane is a very important crop in Cuba. Some people work on plantations, harvesting the plants. Others work in processing factories where the sugar is extracted from the plants and packed, ready for shipping overseas.

▲ The slaughterhouse on the Danish island of Samsø provides jobs for many people.

Cash crops provide jobs and bring money into the islands, but producing them can cause problems. Wages for farm laborers are usually low. Islands that rely on one type of crop for most of their income suffer if the crop is destroyed by disease or bad weather. In 1995, for example, hurricanes destroyed almost all of Dominica's banana crop. Island peoples are trying to develop a wider range of businesses, so that they are not so badly affected by these problems.

From Pigs to Pork

On the small Danish island of Samsø, pigs provide work for many people. Over 100,000 pigs are killed in the local slaughterhouse every year, and over half of these animals are raised by farmers on the island. The slaughterhouse employs 100 people, but many other local store owners and traders depend on the business brought by its workers. Without the slaughterhouse, many islanders would probably be forced to move to the mainland to find work.

▲ Many people in the Greek islands own small boats and earn their living by fishing. These fishermen are at work on the island of Samos.

Fishing

Fishing has always been important work for islanders. Often, people catch just enough to feed their families or to sell on the island. On Tobago in the Caribbean, for example, people go out in small boats or canoes to catch fish in the shallow waters around the coral reefs. In the Caribbean, fish stocks are not big enough for a large-scale fishing industry to develop, and the coral reefs around some islands make it difficult to use large boats.

Where islands are surrounded by seas that are rich in fish, islanders are able to sell fish for export. The people of Newfoundland, in Canada, have traditionally earned their living by fishing for cod in the cold waters of the North Atlantic. Fishermen from Iceland and Japan spend many weeks on large trawlers far out at sea, freezing the fish as soon as they catch it to preserve it until they return to port. Back on shore, fish-processing factories provide jobs for many more people.

Forestry

Logging companies provide jobs on rain forest islands in the Pacific, such as Borneo, New Guinea, and the Philippines. Workers cut down rain forest trees, using powerful chain saws, and transport the huge logs to ports, ready to be shipped abroad. Some island governments are concerned about the number of trees that are being cut down. So they are trying to develop manufacturing industries based on lumber, such as furniture making. More money can be earned from exporting manufactured goods than from exporting logs. This might mean that fewer trees will have to be cut down.

The Forests of Sarawak

In Sarawak on the island of Borneo, logging work has frightened away the wild boar that local people used to hunt. Rivers have become polluted and many fish have died. The local people have protested about the destruction of their forest home. The government has now set limits on the number of trees that can be cut down. It is also using satellites to check that trees are not being cut down illegally.

▼ In the rain forest of Borneo, workers number huge logs, ready to be moved to a port.

ALBUQUERQUE

Mining in Jamaica

Jamaica is the world's third-largest producer of bauxite. Bauxite is used in the production of aluminum, a lightweight metal that is used to make cars and many other goods. Most of the mineral deposits lie quite close to the surface, so miners use explosives to loosen the surrounding rocks and then dig huge holes in the ground to reach the bauxite.

Mining and drilling

Some islanders work in the mining and drilling industries. There are large copper mines in Papua New Guinea and in the Philippines. Cuba is one of the world's biggest producers of nickel. In Trinidad, workers operate oil wells as well as drilling rigs out at sea, which pump up oil and natural gas. Many other people work in the island's refineries, where petroleum products are made. In Great Britain, the oil industry has brought many jobs to northern Scotland and the Shetland Isles.

▼ A miner at a copper mine in Papua New Guinea waits for explosives to blast away the rocks.

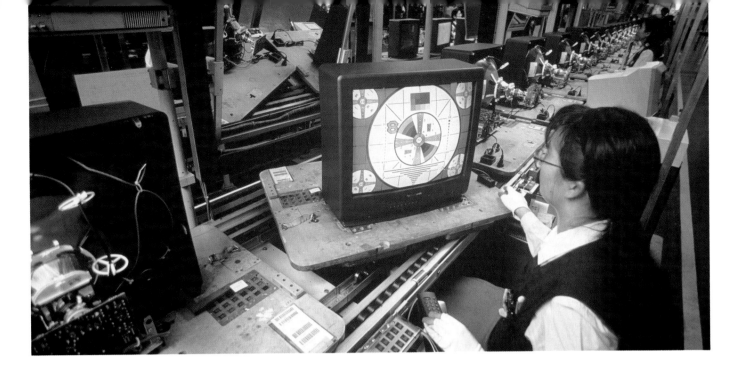

Manufacturing and trade

Islanders in many parts of the world work in manufacturing industries. The textile industry provides jobs in Indonesia and other islands in the Pacific. Chemicals, electronic goods, and medical equipment are produced on the Caribbean island of Puerto Rico. Manufacturing has been important in Great Britain since the nineteenth century. Then, the country was able to use raw materials from its many colonies, as well as its own resources, such as the coal that powered factories. Good transportation links and enough land for industrial development are also important in helping countries to develop manufacturing.

The Japanese government spent a lot of money after World War II (1939–1945) to develop the country's industries. Japan is now one of the most important producers of manufactured goods in the world. More than 5 million people work in the automobile industry in Japan, which produces more cars, trucks, and buses than any other country. The electronics industry makes millions of televisions, computers, and calculators every year. Many factory workers in other island countries, such as Great Britain and Taiwan, work for Japanese companies.

▲ A factory worker in Osaka, Japan, checks the quality of new television sets. Japan's electronics industry is famous throughout the world.

Island Tourism

As air travel has become cheaper and faster, tourism has grown. It is now a vital industry for many island peoples. Tourism is important on islands that are too isolated or are without the resources to develop manufacturing industries or large-scale farming. Beautiful beaches, spectacular scenery, or rare wildlife, which attract tourists, may survive on islands like these when they have disappeared elsewhere. Today, even remote island nations such as Tonga in the Pacific have growing tourist industries.

This islander has a "water taxi" on the West Indian ▶ island of St. Lucia. He takes vacationers for trips in his boat.

BEAUTIFUL ISLE

WELCOME

New jobs

Tourism has changed the lives of some island peoples. Many have given up their traditional work as farmers or fishermen and found new jobs. Some work in the construction industry, building new airports, roads, and hotels. Others work in the new resorts that have sprung up along island coastlines, in hotels, bars, and restaurants.

Islanders sometimes act as guides, taking tourists to see rare wildlife or historic sites. Local craftworkers produce traditional goods, which many tourists like to buy as souvenirs. Dancers in traditional dress put on performances to entertain visitors. Some people believe it is only the tourist industry that keeps island traditions like these alive.

▼ On a Tongan island in the Pacific, a woman performs a traditional dance to entertain some tourists.

Problems of tourism

The growth of tourism has provided a new source of income for islanders, but it has created some problems for them, too. Islands such as Majorca in the Balearic Islands, for example, were among the first to develop large vacation resorts. These were not always well planned. Islanders living in small fishing villages found themselves surrounded by huge modern hotels. Services such as water supplies, sewage treatment, and garbage collection were designed to meet the needs of islanders and could not cope with thousands of extra visitors. The beautiful surroundings that had attracted visitors in the first place were spoiled.

Many island peoples now plan tourist development to avoid some of these problems. In the Seychelles, for example, only 4,500 visitors are allowed on the islands at any one time. On the popular Indonesian island of Bali, resorts are now restricted to specific areas, to reduce the impact of tourism on local people.

▼ A crowded beach at Las Palmas, in the Canary Islands

The resort of Nusa Dua, for example, was built in the early 1980s in an area where the soil was too poor for farming. Deep wells were sunk to provide the resort with supplies of fresh water, and developers were not allowed to build high-rise hotels.

▲ This woman is selling souvenirs on the Caribbean island of Martinique. Many islanders depend on the money they earn from tourists.

Making money

Tourism has provided more jobs for islanders, but jobs such as hotel work are usually not well paid. Many are available only during the main holiday season, when resorts need extra workers to cope with a rush of visitors. Local businesses do not make money from resorts like Nusa Dua, because the money that tourists spend there goes to the large international companies that run the resort. Islands that rely on tourism for most of their income suffer badly if visitors decide to spend their vacations elsewhere.

Settlements and Houses

Just as settlements grew on the mainland, island settlements grew up where there were supplies of fresh water or land for farming. Where island peoples depended on fish for food, many villages developed along the coasts. Some islands are very mountainous, or covered with dense forest, and few people are able to live in these areas. On the four main islands of Japan, for example, 90 percent of the people live on the coastal plains, because three fourths of the land is too mountainous for major settlements.

Many settlements grew up around natural harbors, since these provided shelter for fishing boats and places where traders could unload their goods. Today, many of these settlements have become major cities, such as Wellington in New Zealand and Kingston in Jamaica.

▼ The city of Kingston, in Jamaica, grew up around the large, natural harbor. Ships could anchor and unload easily here.

◄ A cloud of smoke and ash rises from the Soufrière Hills volcano on Montserrat.

Living Dangerously

Some volcanic islands are so small that it is difficult for islanders to find safe places to live if a volcano erupts. The Soufrière Hills volcano on the Caribbean island of Montserrat began erupting in 1995. Homes were swallowed up by lava flows and the capital, Plymouth, had to be abandoned. By 1997, many people had decided to leave the island.

Clues to the past

Clues to island history can be found in many towns. In Charlotte Amalie, the capital of the U.S. Virgin Islands, old Danish-style wooden houses still survive. The islands were ruled by Denmark from the 1700s, before being taken over by the United States early in the twentieth century. The old center of Havana, the capital of Cuba, is laid out like a Spanish city, with squares called *plazas*. There are also some fine mansions, which were built for wealthy plantation owners.

▼ The Plaza de la Catedral, in Havana, Cuba. The Spanish-style buildings indicate that the island was once a Spanish colony.

Island dwellings

A great many island people live in rural areas. In the Philippines, for example, 60 percent of the population lives in the countryside. Traditional houses are found more often in country areas. They are usually built of cheap materials, which are available in the local area.

On cold, treeless islands such as Iceland, people traditionally built their houses of stone, with thick walls and a covering of turf to keep out the cold. Modern houses are more likely to be made of reinforced concrete and designed to withstand earthquakes and winter storms. On many islands of the Caribbean and Pacific, islanders use materials such as wood, bamboo, and thatch. In Cuba, for example, they build houses called *bohios* in the countryside. These are made of wood, and the roofs are thatched with palm leaves.

▼ A traditional house on the island of Teraina in the Pacific. The wood and leaves of island trees are often used as building materials.

◀ A Cuban family outside their *bohio*. Country houses like this are made of cheap local materials.

Family homes

On the island of Borneo, some of the Dyak people live in longhouses, where whole families live together. Traditional longhouses are made of wood, with thatched roofs and bamboo floors. They are raised on stilts, which makes them cooler because air can flow underneath. It also helps to keep out wild animals—or water if there is a flood. Rooms called *biliks*, which run the length of the house, open onto a shared living area. Rice and other foods are dried on an open veranda along the outside of the house. Longhouses that are built today are made of modern materials and have bathrooms.

▼ This traditional longhouse, in Indonesia, was built by the Tanjung Dyak people.

▲ Some of Haiti's poorest people live in this slum in the city of Port au Prince.

The growing cities

Island towns and cities are growing, as more people move to them in search of work. The number of people living in rural areas or outlying islands is becoming smaller, and some areas may eventually be uninhabited. In Fiji, for example, 40 percent of the population now lives in the rapidly growing towns. Over half of these people live in just two cities—Nausori and the capital, Suva—on the main island of Viti Levu. On nearby islands, such as Rotuma, the population is falling.

Housing shortages

Providing enough housing for the growing population is a major problem on some islands. Often, there is only a limited amount of land that is flat enough for farming and for housing. As the towns expand, farmland is lost. Cities such as Manila in the Philippines and Kingston in Jamaica are home to large numbers of poor people who live in slums and squatter settlements. Often, these people have moved from the countryside to the cities but have been unable to find jobs. Those who do have jobs earn very little money.

Island governments need money to provide better housing for their poorest people, but poorer countries often have little money to spend. It can be difficult for vacationers sitting on golden beaches to imagine such problems in places that seem to be "island paradises." But these problems affect cities all over the world, and islands are no exception.

▼ Hong Kong is one of the most crowded islands in the world. Land is being reclaimed from the sea to provide more living space.

High-rise in Hong Kong

Hong Kong Island, on the south coast of China, is a major center for banking and trade. Until 1997, the island, part of the mainland, and smaller islands were a British colony. As the city on Hong Kong Island grew, developers were forced to build upward. Today, high-rise apartments, hotels, and offices soar into the sky. Over 6 million people live and work there, many of them crammed onto Hong Kong Island, which is just 30 sq. mi. (78 sq. km).

Transportation and Communications

Modern transportation and communications are vital to island peoples. In the past, islanders relied on ships to bring them supplies and news of what was happening in the rest of the world. The development of radio communications and, later, air travel made a huge difference, especially to people living on remote islands. Satellite and computer technology have made distance and isolation even less of a problem.

Traditional boats

Some islanders still use traditional boats for transportation. On the smaller islands of Indonesia, simple wooden canoes with outriggers are used. They are very similar to those used by the earliest Pacific explorers. These traditional canoes are used for fishing trips and other short trips, while canoes with sails or outboard motors are used for longer voyages.

◀ These fishermen from Flores Island, in Indonesia, have just cast a net from their boat. Their boat has a traditional design, with outriggers for stability.

Modern boats

Ferries are a vital link in island transportation systems, because they can carry cars, buses, trucks, and even trains. High-speed catamarans and huge car ferries link Great Britain with Ireland and the European mainland, and jetfoils skim among the islands of Japan.

Carrying cargo

Today, most people making long journeys travel by air rather than by sea, but islanders still rely on ships to bring them food supplies and other goods. Many island exports are also transported by ship. In the Caribbean, minerals and agricultural products are loaded onto small ships at local island harbors and taken to large, deepwater ports such as San Juan in Puerto Rico and Port of Spain in Trinidad. Here, they are loaded onto oceangoing container ships to be transported to markets around the world.

▼ This ferry carries tourists and supplies between the Greek port of Piraeus and the island of Póros. Islanders all over the world rely on ferries as a cheap form of transportation.

▲ Small aircraft, like this one on Ovalau Island in Fiji, carry mail and supplies as well as passengers. They also allow islanders to reach doctors and hospitals more quickly.

Traveling by air

Since the 1950s, when the jet airplane first came into use, air transportation has brought many benefits to isolated island communities. It has made contact between neighboring islands much easier, and aircraft now "hop" from island to island, carrying passengers and supplies. Even small island nations, such as Fiji in the Pacific Ocean, have their own airlines.

New businesses

Islanders have been able to develop new businesses with the help of air transportation. Goods that would not survive a long sea journey can now be flown in hours to supermarket shelves in Europe or the United States. This has opened up new markets for island products. For example, growers in the Dominican Republic now produce cut flowers for buyers all over the world.

Air transportation can be important as a way for islands lying on international air routes to earn money. For example, the airline Icelandair flies passengers from Europe to the United States via Iceland's capital, Reykjavik. Every day, tons of goods and thousands of passengers pass through huge international airports such as Narita in Japan and Heathrow in England. Regular and efficient air transportation links are especially important for countries like these, which rely on manufacturing and international trade.

▲ If islands do not have airports or airstrips, like this small Fijian island, aircraft with floats can still reach them by landing on water.

Tourism

The biggest change brought by air transportation has been the development of tourism. Until the 1970s, for example, the only way to visit the Seychelles, in the Indian Ocean, was by ship. Few people had the time or money to make such a journey. In 1971, an international airport was built on the main island, Mahe, and smaller airstrips were built on the outlying islands. Now about 100,000 tourists fly to the Seychelles every year.

Island travel

On islands, as on the mainland, most people travel by road. Larger, wealthier islands, such as New Zealand, Great Britain, and Japan, have well-developed road systems and railroad networks. Bridges and tunnels link islands to other islands and to the mainland. The Channel Tunnel links Great Britain to the continent of Europe. In Japan, the Akashi Kaikyo Road Bridge links the islands of Honshu and Shikoku. At almost 6,560 ft. (2,000 m) long, it is the longest suspension bridge in the world. The Philippines has over 11,000 bridges linking its 7,000 islands.

Providing major transportation systems requires a great deal of money, and poorer islands cannot afford them. Without fast and efficient transportation, it is more difficult for island businesses to develop. It can also be difficult to develop transportation that meets the needs of all islanders. Building international airports and ferry terminals, for example, makes it easier for tourists and goods to reach islands, but islanders may prefer that money be spent on improving local bus services.

◄ Industrialized island nations like Japan can afford to build good networks of roads, but traffic jams are a major problem in the crowded cities.

Telecommunications

Developments in telecommunications have helped island peoples to overcome some of the problems caused by isolation. Islanders can carry out business quickly and relatively cheaply, using computer and telephone links. Satellite television and the Internet bring people up-to-date information.

▼ This satellite dish brings people in Kingston, Jamaica, the latest news and entertainment.

▲ Parts from old vehicles are put together to make colorful "jeepneys," like this one. They are used to carry passengers in the Philippines.

Daily Life and Leisure

Food

In the past, island peoples depended on foods that were grown or collected locally. On many of the Pacific Islands, for example, people used to depend on native plants such as breadfruit and coconuts for much of their food. The people of New Guinea still use the soft center of the sago palm to make flour. Fish has always been very important too. The people of Japan eat three times more fish than meat.

Most islanders in the Pacific grow bananas and plants such as corn, pineapples, rice, and tomatoes, which have been brought to the islands from other parts of the world. Food often reflects the history of an island. In Sri Lanka, for example, curry is popular, because many islanders are descended from people who came from northern India.

▼ Locally produced goods are for sale at this market in Grenada, in the West Indies.

Food today

Today, a wide range of imported foods is available to islanders. However, these tend to be expensive, because the cost of transporting them has to be added to the price. This means that locally grown foods are still an important part of people's diets.

Clothes

The clothing that was traditionally worn by islanders also had to be made from materials that were available locally. In the Pacific Islands, skirts were made of grasses, and shells and flowers were worn as decoration. In the cold climate of Greenland, people wore warm clothes made of sealskin. However, imported clothing has become available to more and more people. In Greenland, for example, most people now wear clothes made of lightweight artificial fibers.

An Inuit man on ▶ Baffin Island, in Canada, wearing a coat made of caribou skin.

Religion

Since they were cut off from contact with other peoples for much of their history, many islanders developed religious traditions and beliefs that were unique to their own island or group of islands. Some of these religions still survive. For example, Shintoism is found only in Japan, and 40 percent of the Japanese people follow this religion today.

Over the centuries, different groups of settlers have brought their religions to islands. The early settlers in the Maldives brought their Buddhist faith with them from Sri Lanka. When Arab traders came to the islands, Islam became more important, and today the islanders are Muslims. People on islands such as Jamaica in the Caribbean and the Philippines in the Pacific were converted to Christianity by European settlers.

▲ This girl is taking part in the Carnival celebrations of Port of Spain, Trinidad. Carnival is a festival that was introduced to the West Indies by Catholic settlers from Europe.

Sports and leisure

International travel and television programs have brought new sports and new kinds of entertainment to islanders, and traditional island pastimes are now well known in many other countries. In Japan, traditional sports include sumo wrestling, Aikido, judo, and karate, but many Japanese people also enjoy skiing and golf. Jamaicans are famous for their local music, reggae, which is popular all around the world. The British brought cricket to their island colonies, and the West Indies now produce some of the world's best cricketers. The Pacific Islanders are famous for dances such as the Hawaiian *hula*. The people of the Philippines enjoy a local ball game, called *sipa*, which requires great skill and control, but American basketball is very popular there too.

Island Languages

Isolated islanders have often developed their own language, which is spoken only by themselves and perhaps a few island neighbors. About 1,200 of the world's 3,000 languages are spoken in the Pacific islands. More than 740 languages alone are spoken on the island of New Guinea.

▼ Steel-band music developed in the West Indies. It has been made popular around the world by West Indians who have emigrated to other countries.

The Future

A big issue facing islanders today is the use of natural resources. Islanders are trying to balance the need to earn money from businesses such as tourism, mining, or forestry, with the need to protect island environments.

Disappearing forests

On many islands in the Caribbean, the Indian Ocean, and the Pacific, charcoal is the main source of fuel, especially in the growing cities. Charcoal is produced from wood, and many island trees are being cut down to produce it. Even more trees are cut down by logging companies, especially on the islands of the Pacific. When trees are cut down, soil can be more easily washed away by rain, and it becomes harder to grow crops.

▼ Smoke rises as trees are cleared from the rain forest in Irian Jaya, Indonesia.

Damaged landscapes

Mining and drilling have damaged the landscape of some islands. Island peoples are also concerned about the pollution caused by these industries, and about the conditions in which islanders have to work. A mine in New Caledonia, for example, produces nickel, which is used in the nuclear energy industries of France and Japan. At the mine's smelting plant, the level of radioactive emissions is a thousand times higher than the level that is permitted in France. Workers are concerned that the emissions could damage their health.

▲ A bauxite mine in Jamaica. Mining brings money into the country but it also damages the landscape.

Fire in the Forests

In 1997, forest fires on the island of Borneo brought misery to millions of people in Southeast Asia. Logging companies that leave behind small, fallen trees and undergrowth that burns easily were blamed for the severity of the fires. Smoke combined with other forms of air pollution to form a thick, poisonous smog. Schools and offices were closed, and people stayed indoors as much as possible. Even so, many people died as a result of breathing in the polluted air.

▲ This man from Savii Island in the Pacific, is using a spear to catch fish. He is carrying the fish he has caught in his mouth.

Overfishing

The number of fish in the oceans is falling. This is a problem that affects island peoples particularly badly, because so many of them depend upon fishing. It is also a problem that some island nations, such as Japan, have helped to create. Some fishing nets are so huge that they can catch up to 200,000 fish at a time. Fish are caught before they have a chance to breed, so there are fewer new fish to replace them. International organizations such as the United Nations have imposed strict limits on the number of fish that can be caught. Although this should help fish stocks to recover, it also means that many island fishermen lose their jobs.

Opportunities for all

Many island nations are overcoming the problems caused by isolation and lack of natural resources. Some islands, such as Jamaica, are developing a wide range of industries, so that they are not so reliant on tourism or cash crops for their income. Singapore and Japan have concentrated on producing electronic goods, which require fewer imported materials and can be sold for large profits. They have spent money on education to produce highly skilled workers. But there are some islands where many people still live in poverty. Haiti, for example, is one of the poorest countries in the Western world. Finding ways to reduce poverty and improve education for all islanders is an important challenge for everyone.

More Fish in the Philippines

People on Banton, in the Philippines, have succeeded in building up fish stocks around their island. They have replanted mangrove trees that had been cut down and have put old tires into the shallow waters, to encourage coral to grow. This provides places where fish can breed and develop. As a result, there are more fish for the islanders to catch and sell.

▼ These islanders are living in poor housing in Jakarta, Indonesia.

Glossary

Bamboo A tall grass with strong stems.

Caribou A North American reindeer.

Catamarans Boats with two hulls that are held together side by side (the hull is the main part of the boat).

Colonies Lands that are ruled by people from another country.

Coral Tiny sea creatures that form rock-like shapes in the sea.

Exports Goods that are sold abroad.

Fault A place where the earth's crust is broken.

Imported goods Goods that are brought in from a foreign country.

Jetfoils Fast, light boats that are raised out of the water on "skis."

Lava Rock that has melted and poured out of a volcano.

Mangroves Trees with roots above the ground that grow along tropical coasts.

Manufacturing industries Industries that turn raw materials, such as steel, into goods, such as cars or computers.

Mineral reserves Places where coal, gas, oil, or metals, which can be used in industry, lie under the ground.

Native Belonging to a particular place, instead of being brought in from outside.

Navigation The method of planning and following the route that a ship needs to take to reach a specific place.

Outrigger A frame that helps keep boats stable.

Plains Areas of flatland.

Plantations Large farms where cash crops are grown.

Radioactive emissions Particles released when nuclear material is broken down.

Rain forest Dense forest found in tropical areas where there is heavy rainfall.

Reef A ridge of rock or coral, the top of which lies just under the surface of the water.

Slums Areas of very poor housing.

Smelting plant A place where rock that contains threads of metal is heated until the metal melts and runs out.

Spices The seeds, leaves, or roots of certain plants.

Textiles Cloth.

Tropical Lying between the Tropics of Cancer and Capricorn.

Veranda A covered area along the side of a building.

Further Information

Books to read

Aylesworth, Thomas G. *U.S. Territories and Possessions; Puerto Rico, U.S. Virgin Islands, Guam, American Samoa, Wake, Midway*. New York: Chelsea House, 1992.

Cerullo, Mary M. *Coral Reef: A City That Never Sleeps*. New York: Cobblehill, 1996.

Jacobs, Judy. *Indonesia, a Nation of Islands* (Discovering Our Heritage). Parsippany, NJ: Dillon Press, 1990.

Kristen, Katherine and Kathleen Thompson. *Pacific Islands* (Portrait of America). Austin, TX: Raintree Steck-Vaughn, 1996.

MacDonald, Robert. *Islands of the Pacific Rim and Their People* (People & Places). Austin, TX: Thomson Learning, 1994.

Myers, Christopher A. *Galapagos: Islands of Change*. New York: Hyperion, 1995.

Sayre, April Pulley. *Coral Reef* (Exploring Earth's Biomes). New York: 21st Century, 1996.

Steele, Philip. *Islands* (Geography Detective). Minneapolis, MN: Carolrhoda Books, 1996.

Waterlow, Julia. *Islands* (Habitats). Austin, TX: Thomson Learning, 1995.

Wu, Norbert. *A City Under the Sea: Life in a Coral Reef*. New York: Atheneum, 1996.

Useful addresses

Friends of the Earth
1025 Vermont Avenue NW
Suite 300
Washington, D.C. 20005-6303
(202) 783-7400

Reforest the Earth
2218 Blossomwood Court NW
Olympia, WA 98502

Earth Living Foundation
P.O. Box 188
Hesperus, CO 81326
(970) 385-5500

The World Rainforest Movement
Chapel Row
Chadlington
Oxfordshire OX7 3NA
Tel: 01608 676691

World Wildlife Fund
1250 24th Street NW
P.O. Box 96555
Washington, D.C. 20077-7795

Forest Stewardship Council
RD 1 Box 182
Waterbury, VT 05676
(800) 244-6257

Picture acknowledgments
The publishers would like to thank the following for allowing their photographs to be used in this book: Bridgeman Art Library 12/Stapleton Collection; J Allan Cash 7, 14, 24, 30, 33, 45; Cephas *Cover*/Wellington Webb; Bruce Coleman 5 (top)/Christer Frederiksson, 19/Christer Frederiksson, 20/David R. Austen, 42/Gerald Cubitt; Ecoscene 29 (lower); Eye Ubiquitous *chapter openers*/James Davis Travel Photography (JDTP), 4/JDTP, 11/JDTP, 38/JDTP, 41/Adina Tovy Amsel, 43/David Cumming; Getty Images *Title page*/Sylvain Grandadam, 5 (lower)/David Hiser, 9/Cris Haigh, 10/Christopher Arnesen, 13, 16/Hugh Sitton, 21/Alan Levenson, 22/Richard Elliott, 23/David Hiser, 25/Sylvain Grandadam, 27 (lower)/Donald Nausbaum, 37 (top)/Paul Chesley, 39 (top)/Paul Chesley, 40/Doug Armand, 44/David Hiser; Erik W. Olsson, Denmark 17; Robert Harding 8, 18, 39/Paolo Koch; Impact 27 (top)/Andy Johnstone, 34/Caroline Penn, 35/E. Quemere/Cedri; NHPA 32/Lady Philippa Scott; Topham Picturepoint 28, 31; Wayland Picture Library 26/Howard Davies, 29 (top)/Tony Morrison, 36, 37 (bottom)/Howard Davies.
The map on page 6 is by Peter Bull.

Index

Page numbers in **bold** refer to photographs.